EARTH'S ROCKS IN REVIEW

SEDIMENTARY ROCKS

By Anna McDougal

Enslow PUBLISHING

Please visit our website, www.enslow.com. For a free color catalog of all our high-quality books, call toll free 1-800-398-2504 or fax 1-877-980-4454.

Cataloging-in-Publication Data
Names: McDougal, Anna.
Title: Sedimentary rocks / Anna McDougal.
Description: New York : Enslow Publishing, 2024. | Series: Earth's rocks in review | Includes glossary and index.
Identifiers: ISBN 9781978537972 (pbk.) | ISBN 9781978537989 (library bound) | ISBN 9781978537996 (ebook)
Subjects: LCSH: Sedimentary rocks–Juvenile literature.
Classification: LCC QE471.M425 2024 | DDC 552'.5–dc23

Published in 2024 by
Enslow Publishing
2544 Clinton Street
Buffalo, NY 14224

Portions of this work were originally authored by Frances Nagle and published as *What Are Sedimentary Rocks?* All new material in this edition authored by Anna McDougal.

Designer: Claire Wrazin
Editor: Caitie McAneney

Photo credits: Cover, p. 1 pjhpix/Shutterstock.com; series art (title & heading background shape) cddesign.co/Shutterstock.com; series art (dark stone background) Somchai kong/Shutterstock.com; series art (white stone header background) Madredus/Shutterstock.com; series art (light stone background) hlinjue/Shutterstock.com; series art (learn more stone background) MaraZe/Shutterstock.com; p. 5 Leene/Shutterstock.com; pp. 7, 30 (weathering) www.sandatlas.org/Shutterstock.com; p. 9 VectorMine/Shutterstock.com; p. 11 morrison/Shutterstock.com; pp. 13, 30 (deposition) Natalya Temnaya/Shutterstock.com; p. 15 Fokin Oleg/Shutterstock.com; pp. 15, 18 arrows Elina Li/Shutterstock.com; p. 16 GoodStudio/Shutterstock.com; pp. 17, 30 (lithification) Yes058 Montree Nanta/Shutterstock.com; p. 18 WARUGRAPHIC/Shutterstock.com; p. 19 thomaslabriekl/Shutterstock.com; p. 21 (top) small smiles/Shutterstock.com, (bottom) Aleksandr Pobedimskiy/Shutterstock.com; p. 23 mdd/Shutterstock.com; p. 25 corlaffra/Shutterstock.com; p. 27 German Globetrotter/Shutterstock.com; p. 29 Budimir Jevtic/Shutterstock.com; p. 30 (erosion) Marco Fine/Shutterstock.com, (sediment layers) NOPPHARAT9889/Shutterstock.com, (arrows) Maksym Drozd/Shutterstock.com.

Printed in the United States of America

CPSIA compliance information: Batch #CWENS24: For further information, contact Enslow Publishing at 1-800-398-2504.

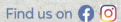

CONTENTS

Words in the glossary appear in **bold** the first time they are used in the text.

EARTH ROCKS!

Earth is made up three kinds of rocks: **igneous**, **metamorphic**, and sedimentary. Igneous and metamorphic rocks make up most of Earth, with sedimentary rocks making up just 1 percent. However, sedimentary rocks make up 75 percent of the land we walk on!

sedimentary rock cliffs

LEARN MORE

Sediment is small pieces of matter, such as stones and sand. Sedimentary rock forms when sediment is **compacted** over a long time.

LET'S BREAK IT DOWN

Weathering is the first step in sedimentary rock creation. Weathering is the breaking down of rock by forces such as wind and water. In **physical** weathering, wind and water move over rock, wearing away pieces of it. This creates sediment!

close-up of sand (sediment)

LEARN MORE

Sedimentary rock can be found all over the world.

Chemical weathering is when water and other matter soften or break down rocks. Both physical and chemical weathering are part of the rock cycle, which is a model that shows how the three kinds of rock form and change.

LEARN MORE

Gases, such as oxygen and carbon dioxide, can have a chemical weathering effect too.

ROCK CYCLE

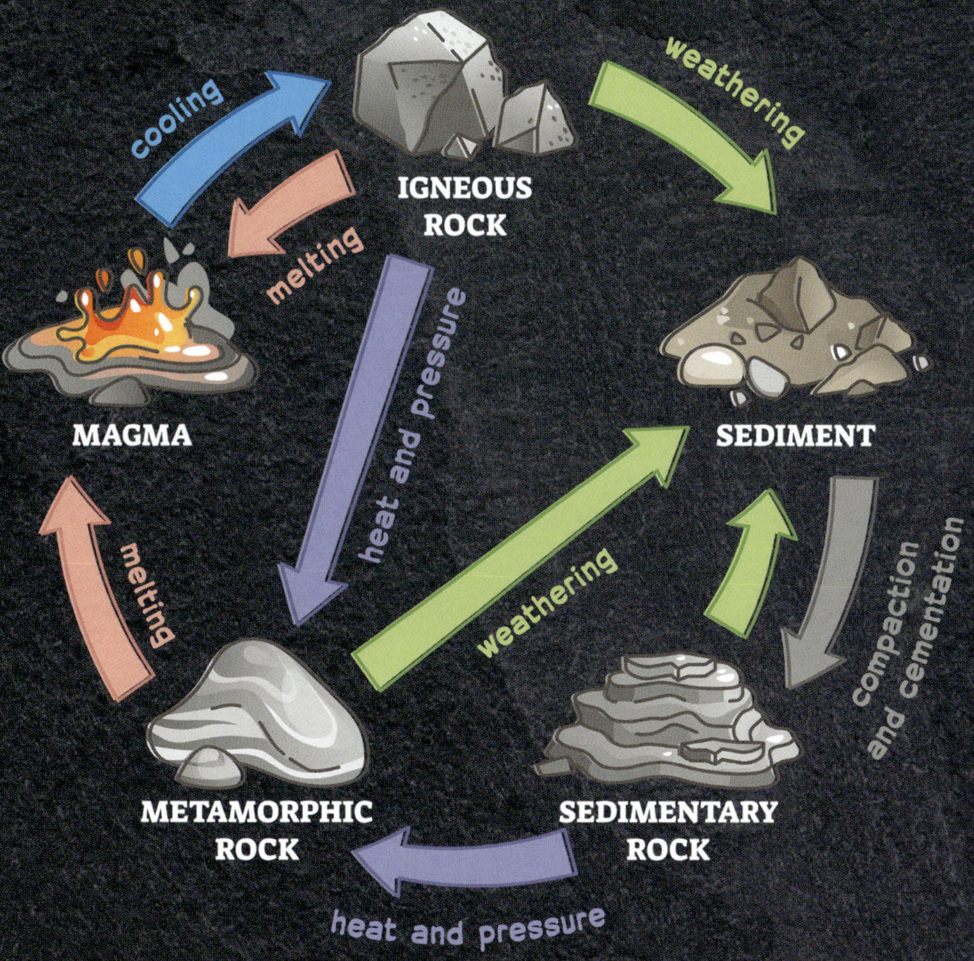
IGNEOUS ROCK
weathering
SEDIMENT
compaction and cementation
SEDIMENTARY ROCK
heat and pressure
METAMORPHIC ROCK
melting
MAGMA
cooling
melting
heat and pressure
weathering

LET'S MOVE IT!

Sediment may not stay in one place forever. It can be carried away by wind or water. It might be pulled down a hill by **gravity**. This movement of sediment is called erosion. Today, erosion is happening faster than ever due to human actions.

rockslide warning sign

LEARN MORE

In rockslides, many rocks fall from a higher place to a lower place, such as from the top to the bottom of a mountain. This is erosion.

During weathering and erosion, bits of sediment often become smaller. They may become round and smooth from bumping and sliding against other sediment. Sometimes, this movement sorts sediment by size or type of rock.

LEARN MORE

Smooth, round pebbles at the beach were likely broken down and moved by water over a very long time to become that size and shape.

COMPACTING SEDIMENT

When sediment lands someplace, it's called deposition. Other matter will build up on top of it. This creates **pressure**. Animals may walk on it, or water may flow over it. That adds more pressure. As pressure increases, the sediment is compacted.

LEARN MORE

Many different kinds of sediment may come together to form a whole new rock when they're compacted together.

The sediments bind together to form a sedimentary rock. This happens when water flows into the open spaces of the rock. **Minerals** that are **dissolved** in the water fill the open spaces. More sediment may settle on top, adding pressure as the sediments bind.

LEARN MORE

The process by which sediments are changed into new rocks is called lithification.

CLASTIC SEDIMENTARY ROCKS

Sedimentary rock comes in two main forms: clastic and nonclastic. Clastic, or detrital, rocks form from the sediment produced by weathering. Clastic rocks sometimes have shells, pebbles, or other large matter in them. These rocks and bits of matter are called clasts.

conglomerate (clastic sedimentary rock)

clastic rock formation

LEARN MORE

Clastic is the most common kind of sedimentary rock. Examples of clastic sedimentary rocks are breccia, conglomerate, sandstone, and shale.

NONCLASTIC SEDIMENTARY ROCKS

Nonclastic sedimentary rocks are formed by lithification, just as clastic rocks are. Nonclastic rocks form in two different ways. Chemical nonclastic rocks are formed from the minerals left behind in a drying area of water.

LEARN MORE

Another kind of nonclastic sedimentary rock is called **organic**. These include the remains of living things such as shells or dead plant matter.

coal (nonclastic rock)

limestone (nonclastic rock)

ROCK LAYERS

Layers in sedimentary rocks are called strata. They sometimes look like stripes of different colors and thicknesses. Strata at the top of sedimentary rock are newer than those at the bottom. Different colors come from the different minerals present.

"The Wave" strata
Coyote Buttes, Arizona

LEARN MORE

The features of a rock's strata depend on where the rock's sediment came from, how much settled there, and what minerals are in it.

The process of forming strata is called stratification. It's also called bedding. Many strata are nearly **horizontal**. Others aren't straight and show the direction wind, water, or gravity moved the sediment as it formed rock. This is called cross-bedding.

cross-bedding in sandstone
Kanab, Utah

LEARN MORE

Some strata have clasts ordered by size, from largest to smallest as you go up. This is called graded bedding.

USING FUELS

We use fossil fuels every day. These fossil fuels, such as oil and gas, are produced from sedimentary rock formation. Organic sedimentary rock that formed from compacted plant matter becomes peat. With enough time, heat, and pressure, peat becomes coal!

LEARN MORE

Both coal and peat can be used for power and heat.

FINDING FOSSILS

Sedimentary rocks tell us about Earth's history. They're the only kind of rock that can hold fossils, or the hardened remains or marks of past plants and animals. Fossils give scientists an important peek into the past.

LEARN MORE

Scientists can learn what life was like on Earth in a certain time period by studying where fossils are in a rock's strata.

FORMING SEDIMENTARY ROCKS

1 WEATHERING

Rocks break down into sediment.

2 EROSION

Sediment is moved.

3 DEPOSITION

Sediment lands in a new place.

4 COMPACTION

Sediment is buried and pressure increases.

5 LITHIFICATION

Sediment is bound together and becomes rock.

GLOSSARY

chemical: Having to do with matter that can be mixed with other matter to cause changes.

compact: To force closer together. Compacted is when something is pressed so that it is harder and fills less space.

dissolved: To be mixed completely into a liquid.

gravity: The force that pulls objects toward Earth's center.

horizontal: Level with the line that seems to form where the earth meets the sky.

igneous: The rock that forms when hot, liquid rock from within Earth rises and cools.

layer: One thickness of something lying over or under another.

metamorphic: Rock that has been changed by temperature, pressure, or other natural forces.

mineral: Matter in the ground that forms rocks.

organic: Having to do with a living thing.

physical: having to do with matter that can be seen or touched.

pressure: A force that pushes on something else.

FOR MORE INFORMATION

BOOKS

Owen, Ruth. *The Rock Cycle.* Minneapolis, MN: Bearport Publishing, 2022.

Rogers, Marie. *Exploring Sedimentary Rocks.* New York, NY: PowerKids Press, 2022.

WEBSITE

Sedimentary Rocks
www.dkfindout.com/us/earth/rocks-and-minerals/sedimentary-rocks/
Check out more fun facts about sedimentary rocks!

Publisher's note to educators and parents: Our editors have carefully reviewed this website to ensure it is suitable for students. Many websites change frequently, however, and we cannot guarantee that a site's future contents will continue to meet our high standards of quality and educational value. Be advised that students should be closely supervised whenever they access the internet.

INDEX